HOW TO GROW BLACKBERRIES
Planting, Growing, and harvesting the bramble fruit

LARRY PAT

TABLE OF CONTENTS

Introduction

If you are contemplating incorporating blackberries into your home garden during this season, you have arrived at the ideal location! Blackberries cultivated in a garden can be just as satisfying and tasty as they are rewarding. Taylor Sievers, a farmer and gardening expert, talks you through each step that you will need to take in order to have a plentiful harvest of blackberries this season in this article.

When you are a gardener, there are very few things that are more satisfying than gathering fruit from your very own backyard. On a lovely summer day, the most enjoyable part is wandering around your garden, picking the fruit off the vine or branch, and savoring the sweet (or sour), juicy, flavor-packed bliss that you have just picked.

You may finish off this picture with a perennial plant that is quite simple to cultivate, and that plant is the blackberry.

There are perennial shrubs that belong to the Rose family, which is called the Rosaceae. Despite the fact that blackberries come in a wide range of shapes, species, and varietals, they are always classified under the Rubus genus. It is possible to cultivate them in virtually any environment, and they are frequently discovered growing wild in temperate regions of the world.

Not only are blackberries delicious, but they also offer a wide range of applications and are simple to store. The needs for their upkeep are rather modest, and they are simple to cultivate. All of the information that you require regarding blackberries and how to properly care for them is covered in this tutorial. Let's get down to business!

Chapter 1

Blackberry Plant Overview

Plant Type: Perennial Shrub
Native Area: North & South America, Europe, Asia
Hardiness Zone: USDA Zones 4 to 8
Season: Summer
Exposure: Full Sun
Maturity Date: 2 years for fruit harvest
Growth Rate: Moderate
Plant Spacing: 4 - 6 feet plants, 8 to 10 feet rows
Planting Depth: 18 inches
Height: 4 to 20 feet
Watering Requirements: Low to Moderate
Pests & Diseases: Japanese Beetles, Orange rust
Tolerance: Temperate
Maintenance: Low once established
Soil Type: Well-Draining
Attracts: Bees
Plant With (Companion): Garlic, Rue, Tansy

Don't Plant With: Conifer trees, wild blackberries
Family: Rosaceae
Genus: Rubus
Species: allegheniensis, argutus, etc.

Plant history

The history of the plant Blackberries have been harvested for its fruit for hundreds of years.

Every plant that belongs to the genus Rubus is considered to be a member of the group known as brambles, which includes blackberries. Raspberries are also called a bramble.

For two thousand years, since the time of the ancient Greeks, blackberries have been harvested and utilized for food and for medical purposes. In Europe, they were harvested from hedgerows and utilized for medical and other reasons until the 16th century. The Evergreen blackberry (Rubus laciniatus) was domesticated in the 17th century in Europe.

The early European inhabitants of North America frequently saw them as a nuisance at first because of their fast growth. The idea of settlers attempting to clear land and having to cope with a tangled and thorny mess is something that comes to mind.

It is possible that early settlers did, in fact, pick fruit from wild stands; but, it was not until somewhere between the years 1850 and 1860 that they were farmed on a large scale in North America. They were produced on around 40,000 acres in the United States by the year 1948. The southeast was the primary location of production.

Following its domestication and further cultivation on a consistent basis, they were eventually classified into three primary categories: upright, semi-erect, and trailing.

The description

They yield fruits that are dark purple and crimson in color, and they have a flavor that is both sweet and sour.

Blackberries are classified as perennial bushes and have evergreen leaves that alternate between three and five leaflets. The leaves are a vibrant green color and have teeth along the margins. Whether or not they have thorns, stems (also known as canes) are semi-woody in texture. This fruit is technically classified as an aggregation fruit since it is made up of several drupelets that have fused together. A drupelet is a small, fleshy fruit that contains a single seed contained within it.

Each aggregate fruit, also known as a berry, is surrounded by a fleshy white cone or core that is referred to as a torus or receptacle. This name is not botanically accurate. During the process of picking the berry, the receptacle becomes detached from the plant.

The receptacle of raspberries is left on the plant after the fruit is removed, which is the primary difference between blackberries and raspberries. Blackberries, on the other hand, are harvested from the plant.

Most species of Rubus are biannual in their natural state. The first season will be spent with them growing in a vegetative state, and the second season will be spent with them flowering, producing fruit, and eventually dying.

Primocanes, also known as the "first" canes, are the terms used to describe the initial canes that are generated following planting. Despite the fact that the leaves will fall off over the winter, the cane will continue to be alive. There is a good chance that you will trim these primocanes in the late winter or early spring, but we shall go into more detail about this topic later.

The transition from spring to summer will result in the growth of new leaves on your primocanes, and ultimately, clusters of tiny white flowers will start to appear between the leaves. As a result of the fact that your primocane is now in its second season, it is classified as a floricane.

You will observe the beginning of the growth of new canes, which can originate from the crown of the plant or from areas that are nearby. These are the primocanes that have arrived on your plant. Suckers are another name for new primocanes that emerge from the root system of your plant in arbitrary locations around the outside of the plant.

Because the floricanes have reached the end of their life cycle, they will eventually perish once they have produced fruit. The primocanes that were produced this year will develop into floricanes the next year, and the cycle will continue this way.

Chapter 2: Care and requirements

Cultivation

Not only do semi-erect variants reach heights of up to 15 feet, but they also lack thrones.

Both the erect and semi-erect varieties of blackberries can trace their roots back to the wild blackberry species that are native to eastern North America.

The height of erect types can range anywhere from six to ten feet (assuming they are not tipped; more on that later). Instead of sucking plentifully from their crown, they sucking abundantly from their roots. Their resistance to cold is rated as moderate to good, and there are cultivars available now that are both thorny and thornless.

Despite the fact that semi-erect types are capable of reaching a height of 15 feet (if not tipped), they generate canes from a crown rather than through a multitude of suckers. They are only able to produce cultivars that are thornless, and their cold resistance is considered to be moderate.

Through a process known as tip layering, arching branches have the potential to develop a new plant by rooting and producing a new plant where they contact the ground.

In contrast to the erect and semi-erect forms, the trailing variants were generated by the breeding of western species. Red raspberries are also in their lineage. Berries of the trailing variety tend to have less apparent seeds and the fruit is more conic in shape.

Additionally, they may ripen quicker with berries that are inherently greater in flavor, sugars, and acid than that of other berries. The canes will form from a crown and grow to a height of ten to twenty feet. They are considered to have a low to moderate level of cold tolerance, and there are varieties that are both thorny and thornless that are available today.

Primocane-fruiting blackberries are a unique variety of blackberry that has been generated from three different types of blackberries.

These will produce fruit in the fall rather than in the summer, and they will do so on canes that have grown during the first (or current) season. This is in contrast to other kinds, which produce fruit on canes that grew during the season before this one. Because of this, it is possible to chop them down to the ground completely throughout the winter.

The process of propagation

The cane will take root and be able to be put in the garden after it has been propagated through the process of tip layering.

Root cuttings, tip layering, and suckers are the three methods that are used to propagate blackberries.

It is possible to take root cuttings during the fall. The roots should be dug up, and you should select those that are at least the size of a lead pencil. Cut them into lengths ranging from three to six inches and keep them in peat moss that is damp at a temperature of thirty-two degrees Fahrenheit. They should be planted mid-winter to early May.

Tip layering is a method that involves bending a cane down to the ground and then adding earth on top of a piece of the cane. Following a period of several weeks, the cane ought to start to develop roots, and you ought to be able to separate it from the rest of the cane in order to make it a separate plant plant. If you do not maintain the canes in an upright position on a trellis, they will naturally tip the layer as well.

It was through the process of tip layering that I began my entire blackberry patch! I proceeded to cut and dig canes from a family member's patch that had naturally rooted into the soil and had arched over the trellis. This was something I did in the middle of spring, and after digging them out, I immediately planted them in my garden.

There is a possibility that suckers will appear close to your patch. In the same way that tip layering works, you can dig up the young suckers and transfer them to your new plant location. When you plant them, you need to make sure that you provide them with plenty of water. During the spring, it is common for "suckers" to appear.

When to plant

The planting of fresh shrubs in the spring is something that is suggested.

When it comes to planting new plants, the best time to do it is in the spring. The primocanes will be able to start growing and the root system of the plant will be able to start establishing itself as a result of this.

With the exception of the plant having an old cane that was left over from the previous season, it is highly unlikely that you will see blossoms or fruit in the first year. Even in such a case, the harvest will be quite sparse, if it occurs at all.

It is also possible to plant in the fall, but you should exercise caution in USDA Hardiness Zones 5 and higher. If you are concerned about the possibility of your young plants being destroyed by freezing temperatures, it is advisable to plant them in the spring.

How to grow

Once they have established themselves, blackberries are quite simple to cultivate. It's possible that the issue you're having is not with developing them but rather with controlling them! Now that you've made the decision to plant them in your garden, let's have a look at the most crucial considerations you need to bear in mind!

Location

These well-known fruiting shrubs thrive best when they are grown in full sunlight.
A large, open location that receives full sun to partial shade is ideal for planting your blackberries. Although your initial plants are very little, they will shortly develop into a dense mass of canes and foliage. Therefore, space is of utmost importance because of this occurrence.

It is not always the case that they are suitable companion plants in the garden due to the rapid growth that they exhibit.

However, there are a few distinct plants with growth patterns that compliment blackberries well; therefore, if you have decided to interplant with blackberry shrubs, you should stay with these companions.

In order to ensure that the plants receive the correct amount of sunlight for growth, full sun is ideal. In full sunlight, plants will produce more fruit and vegetables. Part shade is acceptable for plants in places that are extremely hot, provided that they are receiving afternoon shade.

Not only is it necessary for air flow, but an open site is so. Any increase in air circulation will result in a decrease in the number of cases of sickness.

Spacing

It is recommended that while planting, the bushes be spaced an average of four to seven feet apart.

It is recommended that semi-erect and trailing cultivars be planted in rows that are four to six feet apart in order to guarantee sufficient space. You should plant erect varieties at a distance of three to four feet apart.

Should you have more than one row, you should make sure that each row is at least eight to ten feet apart from the next. You just have to wait till your blackberry patch begins to flourish if you think it is excessive. Later on, you will be grateful to yourself.

It is also possible to plant them in a hedgerow fashion, which essentially means that you are growing the plants in a row closer together so that the plants can use each other as support.

On the other hand, it is quite likely that you will get fewer harvests, and it will be more difficult to harvest from. Planting a hedgerow has a number of benefits, including the fact that it can function as a living fence and requires significantly less effort from the casual gardener.

Soil

Starting with a soil that has a pH level between 5.5 to 6.5 and is well-drained is the first step.

For planting purposes, the ideal soil type is one that has good drainage. However, they are able to withstand soils that are more moist, even more so than raspberries. The ideal pH range for blackberries is between 5.5 and 6.5. It is also recommended to have a significant amount of organic stuff.

Ideal soil organic matter concentration ranges from three to four percent.

To enhance the amount of organic matter that is present in your soil, you can increase the amount of compost or well-rotted manure that you put before planting.

In areas where Phytophthora root rot, Verticillium wilt, or crown gall have been a problem in the past, you should avoid planting in such areas. The planting of tomatoes, peppers, potatoes, melons, strawberries, or weeds such as pigweed, lambsquarters, or weeds belonging to the nightshade family should be avoided in areas that were previously used for growing these types of crops.

Moisture

Drip or trickle watering should be used.
For the success of a berry crop, consistent and enough moisture is of the utmost importance. In the growing season, it is best to irrigate the plant with two inches of water per week.

It is best to water the berries every week at a depth of two to four inches while they are maturing.

Utilizing drip or trickle irrigation is an excellent method for ensuring that your berries receive an adequate quantity of moisture while also minimizing the amount of water that is sprayed on the foliage. Blackberries, on the other hand, are inherently more resilient than raspberries. If there was ever a moment to be concerned about watering, it would be when the new growth and swelling of the berries is just beginning to take place.

Temperature and local climate

The garden is home to the growth of blackberries.

From USDA zone 4 to zone 8, blackberries can be grown.

Every continent, with the exception of Australia and Antarctica, is home to local populations of blackberries.

The USDA Hardiness Zones 4 through 8 are the most common places where they are propagated. The hardiness of certain cultivars may extend up to Zone 3.

To break the dormancy of floricane buds, they need to be chilled for a specific number of hours at temperatures ranging from 25 to 40 degrees Fahrenheit. However, temperatures as low as 10 degrees Fahrenheit might cause damage to the buds of cultivars that are not hardy. Temperatures lower than -15 degrees Fahrenheit have the potential to cause damage to the buds of even the most cold-resistant varieties.

Fertilizer

Fertilizers that contain nitrogen should be used in the springtime, is the recommendation.
If the soil was adequately prepared prior to planting, there is not much cause for concern with regard to the application of fertilizer. Despite the fact that I have never fertilized

my blackberries, they continue to produce a large quantity of juicy berries during each season.

Nitrogen is the one specific nutrient that you might need to be concerned about. Fertilizers that include nitrogen, such as calcium nitrate or ammonium sulfate, can be applied to the soil in the late spring. If you do not choose to make use of a commercially available synthetic fertilizer, you might obtain nitrogen from organic sources such as blood meal and alfalfa meal.

To ensure that your blackberries continue to grow rapidly throughout the year, you can apply a quarter of a pound (0.25 kg) of real nitrogen to each 100-foot row. Alternatively, you could opt to apply a half-rate in the spring, prior to the onset of significant growth, and the remaining half-rate approximately one month before the anticipated harvest. The second treatment for blackberries that bear fruit in the fall should be postponed until late summer.

Mulching

The use of mulch helps to maintain the moisture level of the soil, inhibits the growth of weeds, and stops water from splashing onto the leaves.

Mulching, which is a process that is generally good, can assist in increasing your yield. To begin, mulching helps store moisture and minimizes the amount of moisture that is lost from the soil as a result of evaporation.

Mulching will help lessen the amount of weed pressure that is present around your newly planted areas.

Furthermore, over time, mulch will decompose, which will result in the addition of organic matter to the soil. As the mulch starts to break down, it is important to make sure that you reapply it every few years.

Trellising

You can assist them in climbing off the ground by installing a trellis.

Trellises are going to be necessary for blackberries that are semi-erect and trailing. When you use a trellis, the primary objective is to prevent the plants from growing on the ground. There are many different kinds of trellises, but the primary goal is to keep the plants from growing on the ground.

There are two options available to you: either you may construct a trellis in which each cane must be tied separately, or you can use a trellis in which you can weave the canes through in order to keep them placed. Make use of robust posts that are capable of bearing the weight, and make sure that they are buried deep enough in the ground to prevent frost-heaving before they are used.

Pruning

In the late winter and early spring, prune the trees.

In order to achieve the best possible berry harvests and to protect against disease, pruning is absolutely necessary. The fact that different varieties of blackberries require slightly varied pruning techniques in order to get maximum yield is the challenging aspect.

Pruning should be done between the end of winter and the beginning of spring. Because the plants are in a dormant state at this time, this type of pruning is referred to as dormant pruning. For illustration purposes, my farm is situated in USDA Zone 6 in the middle of the United States, and the best time for me to prune is between the end of February to the middle of March.

An additional method of pruning is referred to as tipping. The act of tipping, which is often performed in the early summer, fosters the development of lateral growth from the canes. In essence, lateral expansion is the formation of side branches.

Erect blackberries pruning

Forms of blackberries that are upright Dormant pruning should be performed on blackberry kinds that stand upright so that the canes are spaced approximately ten inches apart. During the dormant season, you should also prune the laterals, which are classified as side branches, so that they are only approximately 12 to 18 inches in length. To ensure that the canes are only three to four feet tall, the plants should be tipped in the early summer.

Semi-erect blackberries pruning

Blackberries that are only partially upright In the early summer, semi-erect types should be tipped to a height of approximately five feet, and the lateral branches should then be tied up to the trellis in the latter summer months.

When performing dormant pruning, it is important to keep the top five to eight lateral branches rather than cutting off any lateral branches that are located underneath or within the bottom three feet of the main cane.

When trimmed, the laterals should be cut to a length of between 12 and 18 inches. First, you should tie the vertical parts of the canes while they are still green. After that, you can move on to tying the lateral branches on the trellis in a horizontal position.

Trailing blackberries pruning

Primocanes that are trailing trailing blackberries should not be tipped. Because the canes have a tendency to trail on the ground, it is important to lift them and secure them onto a trellis in the appropriate manner. Canes can be raised in the late summer or fall, or they can be left on the ground over the winter. Both activities are possible.

It is stated that those that are tied up in the fall have larger fruit harvests; however, if you live in a location that is colder, you should be aware that you may have frost damage when you are training your plants in the fall seasons. Your canes should be reduced to between six and ten healthy canes. Canes that look to be weak should be removed.

Furthermore, once the floricanes have finished fruiting for all types of berries, with the exception of the primocane-bearing blackberries, and you have collected your berry crop, it is a good practice to remove the floricanes that are withering as soon as possible. In addition to reducing the amount of trimming that needs to be done during the dormant season, this will ensure that the canopy is opened up, which will result in improved air circulation.

Chapter 3

When and how to harvest

They are ready to be harvested once you have observed that they have reached their full color and are simple to pluck through.

At long last, the time has come for you to go out and pick some of those luscious blackberries! You will know that they are ready to be picked when they have reached their full maturity in color and can be removed with relative ease.

Just so you know, when raspberries are ripe, they are easy to remove since the receptacle that is inside of them remains behind when you pull them off. The receptacle is removed along with the remainder of the berry during harvesting, which results in a lower degree of "give" in comparison to the process of harvesting raspberries.

You can also determine whether or not your berries are ready to eat by tasting them.

A sour or flavorless taste is characteristic of unripe fruit. Take a taste test every day until you find the harvest time that works best for you. There are some types that have a naturally sour flavor, so you shouldn't be surprised if you can't seem to locate the best time to harvest them. All you need to do is sprinkle some sugar on them, and you'll be set to go.

I find that the best time to pick is when the berries have reached their full coloration (bluish-black; there is no red) and the texture of the fruit is soft rather than stiff. When the berry is softer, the amount of time it may be stored is shorter; however,
I typically consume or store them straight away.

Commonly Used and popular Varieties

1. Marion blackberry
The 'Marion' cultivar is a thorny one that produces fruits that are aromatic and delicious.

2. "Navaho" is a variety that, due to its upright growth, is excellent for cultivation in the northeast, southeast, midwest, and northwest regions. There are no thorns on the 'Navaho' plant, and the fruit ranges in size from tiny to medium. Because of its resistance to anthracnose and root rot, this cultivar is recommended.

3. The erect variety known as **"Kiowa"** is a mid-season producer that is ideal for the southeast and midwestern regions of the United States. In addition to having enormous, tasty fruit, this thorny cultivar also has a lengthy ripening season.

4. The 'Prime-Ark 45' variety is an erect variation that bears fall berries and primocanes. It is a late season variety. Firm and of a medium size, the berries are also firm. This particular cultivar might not ripen in time for a satisfactory harvest in certain regions.

5. "Chester Thornless" is a late-ripening cultivar that is semi-erect and thornless. It produces berries that are huge, dark black, and deliciously sour. Furthermore, it is superior in terms of cold hardiness and yield in comparison to the comparable cultivar known as "Hull Thornless."

6. "Hull Thornless" is a thornless variety that bears fruit in the middle of the season and is considered to be one of the sweeter and less tart varieties of the semi-erect kind.

7. The 'Triple Crown' variety is a semi-erect variety that bears late and does not generate thorns. However, it is a plant that is noted for its high level of vigor, producing huge berries that have a flavor that is extremely sweet.

8. The thorny trailing variety known as **"Marion"** has been a mainstay in the Northwest for a very long time. It is common practice to refer to this cultivar as "marionberry" when it is sold. The fruit is of a medium size, has a pleasant aroma, and is exceptionally tasty. The texture of fruit might be mushy and uneven.

9. The trailing variation known as **"Columbia Star"** is thornless and ripens between the beginning and the middle of the season. The bushes are robust and produce large, consistent berries that have a flavor that is rather exceptional.

10. **'Logan'** is a trailing thorny or thornless type that matures early and has berries that range in size from medium to large and are a raspberry red color.

Chapter 4: Disease and pests

Fungal disease

Despite the fact that blackberries are quite resistant to the majority of diseases, there are still a few that you should be on the lookout for. Let's have a look at some of the most typical fungal issues that you could experience while cultivating plants in your yard. What are they?

Gray Mold (Botrytis Fruit Rot)

Botrytis fruit rot, often known as gray mold The disease known as Botrytis Fruit Rot is distinguished by the presence of a covering on the berries that is whitish-gray in color.

The pathogen Botrytis is typically responsible for this issue, which manifests itself after the harvest of berries and occurs when the berries are stored in the refrigerator. Within the berries, you will observe a growth that is a whitish-gray color.

There is a possibility that the sick berries could get so overgrown with gray fungal growth that they will look to be mummified if they are allowed to remain on the plant. Increasing air circulation by trellising, training, and trimming to ensure that there is adequate space between canes is the most effective method for lowering the risk of having an infection.

The anthracnose

Anthracnose is a disease that causes the canes of the plant to die and it manifests itself as purple spots on the leaves of the plant.

The infected leaves and canes will acquire patches that are a reddish color and will become larger. Over time, the gray cores of the spots on the leaves may disappear, and the spots on the canes may become sunken. Both of these changes may occur. Stunting and girdling of the canes will ultimately result in the death of a significant number of canes.

It is recommended to select types that are resistant to anthracnose and/or to apply lime-sulfur or copper fungicides once during the springtime, when the new leaves are just beginning to emerge. Be cautious to remove and burn any canes that are affected with the disease.

Orange rust

If the leaves or the entire plant are affected, it is suggested that they be removed.

During the spring, this sickness can be clearly diagnosed. In addition, the young shoots will develop into spindly and feeble shoots, and the leaves will be small and pale green to yellowish in color. Over the course of a few weeks, blisters will develop on the leaves, and when they burst, they will discharge masses of rust-colored fungal spores that are powdery in appearance and spread by wind.

At the end of spring, the leaves will begin to wilt and fall off. Despite the fact that the canes will continue to be infected and will not produce blossoms in the future, the new leaves that grow at the tips will appear normal. Immediately eliminate the sources of infection by removing plants in their whole body before the fungal spores are released into the environment. When you are planting new plants, you should make sure to get them from a nursery that has a good reputation.

Insects pest

Unfortunately, blackberries are not immune to having pests attack them. There are a number of insect pests that are capable of causing considerable damage to the plants that you garden. Let's take a look at the ones that are the most prevalent, as well as the methods that may be used to manage them.

Japanese Beetles
The plant leaves are the food source for these insects.
The appearance of these beetles often occurs in the middle of the season.

They have bodies that are metallic reddish-brown in color and glossy. In a relatively short amount of time, they are able to essentially skeletonize bramble leaves, which they consume as a source of nutrition.

On the other hand, plants that are between one and three years old are more susceptible to damage than established plantings, which may not be affected as much. It is possible to acquire commercial traps; however, it is important to note that these traps are known to attract Japanese beetles. Therefore, if you intend to use these traps, set them a considerable distance away from your plants.

Another alternative to using insecticides is to manually pick beetles every one to two days and place them in a pail of soapy water. This is an alternative to using insecticides. As a result of the soap, the beetles are unable to climb up the side of the bucket, and as a consequence, they end up drowning in the water.

If you are able to maintain control of it, this has the potential to be incredibly effective.

Additionally, it is possible that planting companion plants such as garlic, tansy, or rue will be useful in warding off Japanese beetles. Beetles are drawn to plants that have been harmed, thus it is best to prevent damage as early as possible.

Having Spotted Wings Eggs are laid by drosophila in plants that are still growing.
The appearance of adult flies is strikingly similar to that of fruit flies; however, male flies are distinguished by a big black spot that is located on their wings. Located inside the fruit, the larvae have the appearance of very small white worms and can be harvested.

If you have picked blackberries directly off the plant and consumed them, it is highly probable that you have also consumed a few of these minuscule larvae.

When the fruit is submerged in a warm saltwater solution, the larvae will float to the surface of the water.

Adult flies will lay their eggs in the fruit that is still developing, which will result in the formation of a small scar around the fruit. This scar may eventually cause the fruit to collapse and decay.

Another option is to immediately place the berries in the refrigerator in order to prevent the larvae from developing any further. This ought to be done in addition to the use of pesticides, which should only be done if you KNOW the larvae are there and you know that it will be an issue for you. Leaving overripe fruit on the plant or berries on the ground is to be avoided at all costs. Get rid of any debris that may be in the vicinity.

The blackberry Psyllids consume leaves, which causes the leaves that have been harmed to curl.

Psyllids are little winged insects that have stripes on their wings that are a reddish-brown color. They are similar to aphids. They will jump once they are disturbed. In the process of feeding on the leaves, adults will leave behind a milky white material.

The leaves will curl because of the severe damage that has been done. Damage caused by psyllids will typically not have a significant impact on yield, which is why pesticide treatment is not suggested. Planting them in close proximity to a stand of conifers is the most effective method of prevention since they like to spend the winter in pine trees.

Cane borers

Shoots that have been infested with cane borer are either extremely feeble or completely rot away.Cane Borers By digging into the canes of brambles, cane borers, including red-necked, flat-headed, and bronze cane borers, can be considered pests that cause damage to brambles. This causes swellings to grow at the bore sites on canes, which are often not more than one foot above the ground.

Boring sites can sometimes be observed from a height of up to four feet above the earth in some situations. Canes that are infested will either perish or suffer serious damage. If you come across any canes that are sick, you should promptly prune them out and burn them. Prior to the blooming stage, insecticides can be administered.

Yellowjacket wasps
The presence of ripe fruits is what attracts yellow jackets.
They may provide a risk to persons who are picking the berries, even though they may not necessarily pose a harm to the fruit themselves. As a result of the fact that yellowjackets are drawn to fully ripe and damaged fruit, they have the potential to pose a threat to fruit pickers who are not paying attention.

By utilizing sufficient trellising and support, you can keep maturing berries off the ground, and you should also select ripe fruit on a regular basis. Another option is to place traps around the border of the patch prior to the ripening of the berries.

The picnic and the sap beetles
For the purpose of laying their eggs, these beetles will enter ripe fruits and lay their eggs there.
In order to deposit their eggs, these beetles will bore into completely ripe and overripe fruit. As a result, the fruit will be damaged, and there is a possibility that the fruit will become infected with a disease caused by fungus that causes fruit rotting. Additionally, they have the potential to become a contamination in fruits that have been picked.

It is possible to attract and capture insects that may be looking for a sweet treat by selecting fruit that is broken or overripe and placing it in a plastic bag that is located away from the area. After the Beatles have been captured, the bag should be tied and then destroyed in order to get rid of part of the bugs.

Chapter 5: Preserving

If you want to extend the shelf life of blackberries, it is recommended that you freeze them.

Berries should be refrigerated as soon as possible for as long as possible. It is recommended that you bring the berries inside in shifts as you pick them if you are harvesting on a hot day and you have a significant harvest.

Wait until you are ready to use the fruit before washing it. When you do wash the berries, make sure to handle them with care. After washing the berries, use a paper towel to absorb any moisture that may be present before placing them in the refrigerator.

As a result of the fact that they do not have a very lengthy shelf life, it is a good idea to freeze them if you are aware that you will not be using them right away.

There is also the possibility of preparing delectable jams, jellies, and preserves with blackberries.

Uses of Plants

Utilize them in the preparation of jam, pies, or ice cream.

On a hot summer day, blackberries are a delicious sweet delicacy that you should definitely try. When I am in the evening, my favorite thing to do is to snack on them while I am cutting flowers in my cutting garden.

Blackberries can be consumed if they are fresh or if they are drizzled with sugar. Throughout the summer, a great number of pies and cobblers have been prepared with fresh fruits. Additionally, they can be preserved in the form of jams, jellies, and preserves.

As an additional alternative, you may turn them into brandy or wine.

The flavoring of liqueurs and cordials is another application for them. Antioxidants, potassium, phosphorus, iron, and vitamins A, C, and E are all found in extremely high concentrations in blackberries.

It is also possible to use their leaves and canes, which contain unripe berries, as filler and greenery in flower arrangements that have been taken away. Considering that I am a farmer who grows cut flowers, suckers are an essential component of the early spring bouquets that I create.

Chapter 6 : 21 delicious blackberry recipes for beginners

1. Blackberries smoothies

Ingredients: - One cup of blackberries that are fresh
1. One banana
- One cup of almond juice
Honey, one tablespoon's worth
- One-half cup of Greek yogurt
Details to follow:
To begin, put all of the ingredients into a blender.
2. Blend until it is completely smooth.
3. Transfer the mixture to a glass and serve it right away.

2. Blackberry cobbler

Ingredient
- Four cups of fresh blackberries
1 cup of sugar, 1 cup of flour that rises on its own, and 1 cup of milk
- One-half cup of salted butter

Details to follow:

1. The oven should be heated to 175 degrees Celsius (350 degrees Fahrenheit).

2. Arrange the blackberries in a baking dish that has been buttered and then sprinkle them with sugar.

3. In a bowl, combine the flour, milk, and butter that has been melted. Pour the liquid over the berries.

4. Bake for forty-five to fifty minutes, or until the top is golden brown.

3. Blackberry jam

Ingredient:

four cups of fresh blackberries, three cups of sugar, and one tablespoon of lemon juice

Details to follow:

In a large pot, mash the blackberries until they are smooth.

Add the sugar and lemon juice, and give it a good swirl.

3. While stirring constantly, bring to a boil and continue cooking for fifteen to twenty minutes.

4. Pour the mixture into jars that have been sterilized and then seal them.

4. Blackberries salad

Ingredients: - Four cups of a variety of greens
1 cup of blackberries that are fresh
1/2 cup of crumbled feta cheese and 1/4 cup of almonds that have been sliced
- One-fourth amount of balsamic vinaigrette
Details to follow:
1. Put the blended greens, blackberries, feta cheese, and almonds into a big bowl and mix them together.
Secondly, drizzle the salad with balsamic vinaigrette and gently toss it to coat it.

5. Blackberry muffins

Ingredient:
Two cups of all-purpose flour, half a cup of sugar, and two teaspoons of baking powder
- One cup of milk - One-half of a peppercorn
- One-fourth cup of melt butter

a single egg and one cup of fresh blackberries

Details to follow:

1. The oven should be heated to 375 degrees Fahrenheit (190 degrees Celsius).

Put the flour, sugar, baking powder, and salt into a bowl and mix them together.

The third step is to combine the milk, melted butter, and egg in a separate bowl. Mix in the dry ingredients until they are almost completely incorporated.

4. Fold the blackberries in with tenderness.

5. Fill muffin cups to a level that is two-thirds full, and bake for twenty-five to twenty-five minutes.

6. Blackberry Lemonade

Ingredient:

Two cups of fresh blackberries and one cup of lemon juice

Four cups of water, Half a cup of sugar

Details to follow:

To remove the seeds, first blend the blackberries and then filter them.

2. Put the pureed blackberries, lemon juice, sugar, and water into a pitcher and mix them together.

3. Give it a thorough stir, then place it in the refrigerator until it is completely cooled.

7. Blackberry pancakes

Ingredient:

one cup of all-purpose flour, two tablespoons of sugar, and one tablespoon of baking powder

- One cup of milk - One-half of a peppercorn

- One egg - Two tablespoons of butter already melted

1 cup of blackberries that are fresh

Details to follow:

To begin, combine the flour, sugar, baking powder, and salt in a bowl first.

2. With a whisk, combine the milk, egg, and melted butter in a separate bowl.

3. Mix together the liquid and dry ingredients, then incorporate the blackberries.

4. Cook the pancakes on a griddle that has been buttered until bubbles appear, then turn them and continue cooking until they are golden brown.

8. Blackberry Ice Cream

Ingredient
2 cups of blackberries that are fresh
One cup of sugar
Two cups of heavy cream
A single cup of whole milk
1 teaspoon of essence of vanilla bean
Details to follow:
1. Remove the seeds from the blackberries by straining them after they have been pureed.
Blackberry puree, sugar, heavy cream, milk, and vanilla extract should be combined in a bowl and stirred together.
3. Pour the mixture into an ice cream maker and churn it in accordance with the instructions provided by the manufacturer.

9. Blackberry sauce

Ingredient

Two cups of fresh blackberries, half a cup of sugar, and one tablespoon of lemon juice

a mixture of one tablespoon of cornstarch and two tablespoons of water combined

Details to follow:

1. Put the blackberries, sugar, and lemon juice into a saucepan and mix them together.

The second step is to cook the blackberries over medium heat until their juices are released.

3. Pour in the cornstarch mixture and whisk gently until it becomes thick.

4. Place on top of sweets or pancakes while still warm.

10. blackberry scones

Ingredient

Two cups of all-purpose flour, one-fourth cup of sugar, and one tablespoon of baking powder

Half a teaspoon of salt and half a cup of chilled butter, cut into cubes

- One-half cup of thick cream

1. One egg

1 cup of blackberries that are fresh

Details to follow:

1. The oven should be heated to 400 degrees Fahrenheit (200 degrees Celsius).

Put the flour, sugar, baking powder, and salt into a bowl and mix them together.

3. Integrate the butter into the mixture until it resembles coarse crumbs.

4. In a separate bowl, combine the cream and the egg by whisking them together. Mix in the dry ingredients until they are almost completely incorporated.

5. Gently incorporate the blackberries into the mixture.

Sixth, place the dough on a baking sheet and bake it for fifteen to twenty minutes.

11. Blackberry Pie

Ingredient

four cups of fresh blackberries, one cup of sugar, one-fourth of a cup of cornstarch, and one tablespoon of concentrated lemon juice
- One crust for the pie

Details to follow:

1. The oven should be heated to 375 degrees Fahrenheit (190 degrees Celsius).

2. Combine the blackberries, sugar, cornstarch, and lemon juice in a basin that is clean.

After pouring the mixture into the pie crust, cover it with another crust or a lattice.

4. Bake for forty-five to fifty minutes, or until the top is golden brown.

12. Blackberries palette

Ingredient

Two cups of fresh blackberries, one-fourth of a cup of sugar, one tablespoon of cornstarch, and one pie crust are the ingredients.

Details to follow:

1. The oven should be heated to 400 degrees Fahrenheit (200 degrees Celsius).

2. Combine the blackberries, sugar, and cornstarch in a well-disposed basin.

3. Place the blackberry mixture in the middle of the pie crust that you have rolled out.

4. Fold under the edges of the filling, and bake for twenty-five to thirty minutes.

13. Blackberry Parfait

Ingredient

- Two cups of Greek yogurt

1 cup of blackberries that are fresh

- A half cup of granola

Honey, two tablespoons' worth

Details to follow:

Using a glass, layer granola, blackberries, and yogurt in a layered fashion.

2. Serve the dish immediately after drizzling it with honey.

14. Blackberries cheesecake

Ingredients:

One cup of crumbs made from graham crackers

- One-fourth cup of melt butter
- Cream cheese, 16 ounces amount, softened

1 cup of sugar, 1 teaspoon of vanilla extract, and 2 eggs are required.

1 cup of blackberries that are fresh

Details to follow:

1. The oven should be heated to 325 degrees Fahrenheit (160 degrees Celsius).

3. Combine graham cracker crumbs and butter that has been melted. Using a springform pan, press into the bottom of the pan.

3. Begin by beating the cream cheese, sugar, and vanilla extract together in a bowl. The egg should be added in turn.

Pour the sauce over the crust, and then sprinkle the blackberries on top.

Bake for forty-five to fifty minutes.

15. Blackberry Sorbet:

Ingredient

- Four cups of completely fresh blackberries
One cup of sugar, half a cup of water, and one tablespoon of lemon juice together

Details to follow:

To remove the seeds, first puree the blackberries and then strain them.

2. Sugar and water should be mixed together in a pot. Sugar should be dissolved by heating.

3. Combine the pureed blackberries with the syrup and the lemon juice.

4. Defrost the mixture in an ice cream maker in accordance with the instructions provided by the manufacturer.

16. Blackberry Vinaigrette:

Ingredient

- One cup of fresh blackberries
It is recommended to use a quarter cup of olive oil, two tablespoons of balsamic vinegar, one spoonful of honey, and salt and pepper to taste.

Details to follow:

1. Puree the blackberries until they are liquid.

To the mixture, add honey, balsamic vinegar, olive oil, and seasonings of your choice.

3. Blend until everything is just right, and then serve on top of salad.

17. Blackberry crumble

Ingredients: four cups of fresh blackberries, half a cup of sugar, and one tablespoon of lemon juice

• One cup of all-purpose flour • One half cup of oats

- One-half cup of white sugar

1 and a half cups of melted butter

Details to follow:

1. The oven should be heated to 175 degrees Celsius (350 degrees Fahrenheit).

2. Combine blackberries, sugar, and lemon juice in a bowl and stir properly. To bake, place in a baking dish.

3. In a separate bowl, combine the flour, oats, brown sugar, and butter that has been melted. On top of the blackberries, sprinkle.
4. Bake for thirty to thirty-five minutes.

18. Blackberries mojito

Ingredient
- One-half cup of chopped fresh blackberries
- Eight mint leaves - One tablespoon of brown sugar
- The juice of one lime
- Two or three ounces of white rum - Club soda

Details to follow:
First, in a glass, muddle together the blackberries, sugar, and mint leaves.
2. Mix together some white rum and lime juice.
3. Place ice in the glass, and then pour club soda on top of the ice. Give it a good stir.

19. Blackberry bars

Ingredient

One cup of all-purpose flour and half a cup of oats are the ingredients.

- One-half cup of white sugar

1 and a half cups of melted butter

2 cups of blackberries that are fresh

Sugar, one-fourth of a cup, and cornstarch, one tbsp

Details to follow:

1. The oven should be heated to 175 degrees Celsius (350 degrees Fahrenheit).

2. Combine melted butter, flour, oats, and brown sugar in a mixing bowl. Use a baking dish to press the other half.

3. Pour the sugar, cornstarch, and blackberries into a bowl and stir well. Spread on top of the crust.

4. Using the leftover crust mixture, sprinkle it on top.

5. Bake for thirty to thirty-five minutes.

20. Blackberry Glazed Chicken

Ingredient

- Four boneless, skinless chicken breasts

1 cup of blackberries that are fresh

In addition to two teaspoons of honey, a quarter cup of balsamic vinegar, and salt and pepper to taste

Details to follow:

1. The oven should be heated to 375 degrees Fahrenheit (190 degrees Celsius).

The chicken should be seasoned with salt and pepper before being placed in a roasting dish.

3. Blend the blackberries, honey, and balsamic vinegar together in a blender until smooth.

On top of the chicken, pour the mixture.

5. Bake for twenty-five to thirty minutes, or until the chicken is all the way done.

21. Blackberry Cake

Ingredient

- One cup of sugar - Half a cup of butter, melted
-- Two eggs
1 teaspoon of essence of vanilla bean
— One and a half cups of all-purpose flour
One and a half tablespoons of baking powder
- Half a cup of milk
- One and a half cups of fresh black fruits

Details to follow:

1. The oven should be heated to 175 degrees Celsius (350 degrees Fahrenheit).
2. Melt the butter and sugar together in a mixer. Eggs and vanilla essence should be added.
3. While in a separate basin, combine the flour and baking powder. Add the butter mixture in increments, alternating with milk as you do so.
4. Fold the blackberries in with tenderness.
5. Pour into a cake pan that has been oiled and bake for thirty to thirty-five minutes.

Conclusion

Blackberries are the greatest option if you are trying to add fruit to your garden because they are simple to cultivate, they are rich in antioxidants and flavor, and they are fruitful even for the most casual or starting gardener.

They are a perennial crop that is productive and will continue to produce for a number of years. It is possible that you may discover that the most challenging aspect of producing blackberries is the task of managing these unruly brambles! I guarantee that the wait will be well worth it, despite the fact that it takes them around two years to become established.